LOOKING AT SCIENCE 2

The Natural World

David Fielding

Basil Blackwell

© 1984 David Fielding

All rights reserved. No part of this publication may be reproduced, stored in a retrieval system, or transmitted in any form or by any means, electronic, mechanical, photocopying, recording or otherwise, without prior permission of Basil Blackwell Publisher Limited.

First published 1984

Published by Basil Blackwell Limited
108 Cowley Road
Oxford OX4 1JF

ISBN 0 631 91360 2

Printed in Hong Kong

Topic symbols

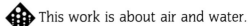 This work is about air and water.

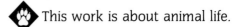 This work is about animal life.

◆ This work is about electricity and magnetism.

◆ This work is about light and dark.

◆ This work is about mechanics.

◆ This work is about plant life.

◆ This work is about weather and climate.

Look for the symbols in the other books in the series. There is more work about these things in the other books.

Contents

A word to teachers and parents 5

Part 1   Nature

Fossils 6
What fossils are and how they are formed

Mud and soil 8
How soil is made; the different kinds

Plants 10
How plants began, how they changed and what they need

Roots and stems 12
How they work for the plant

Leaves 14
Why plants need leaves; what they do

Flowers at work 16
How pollen is made and spread

Summary: The world 18

Part 2   Animals

An animal hunt 20
Looking at small creatures

Animal traps 22
Small creatures and their food

Bones 24
What they are for; why some animals have them

How bones move 26
Joints and muscles

Flying 28
The importance of shape in flight

Life in water 30
Why fish have a special shape

Summary: The animal world

Part 3   Mysteries

Walking on water 34
Why something that should sink can rest on water

On the level 36
How water will always make one flat surface if it can; the siphon

Can you believe your eyes? 38
How light travels, and how certain things can bend it

Moonlight 40
Why the moon shines and changes shape

Rainbows 42
How rainbows are formed

Lightning and other sparks 44
What lightning is and how it is caused

Summary: Explorers of mysteries 46
New words 48

Acknowledgements

Barry Angell   17, 43
Alan Beaumont   23, 28, 31(4), 37,
   cover (front and back)
Biofotos   31(5), 34
Janet and Colin Bord   11(l.)
David Collins   11(r.), 23(2), 33
Edward Hankey   27
National Coal Board   18
Frank Lane   45(4)
Ken Pilsbury   45(5)
Rida Photo Library   6
Flip Schulke/Seaphot Limited:
   Planet Earth Pictures   39
ZEFA   40, 42
The Zoological Society of London   25, 27

Illustrations by Michael Stringer (colour)
and David Fielding (black and white)
Design by Indent, Reading

A word to teachers and parents

This is the second book in the *Looking at Science* series, which sets out to do two things:
- To give children a body of basic knowledge in natural and physical science;
- To introduce them to the nature of scientific enquiry.

These elements are developed side by side throughout the five books.

Each double page covers a particular area of study, with information and activities. The activities are introduced with the symbol ♥, and cover experimentation, observation and recording. A list of all the equipment needed for the experiments is given near the beginning of each spread.

The books can be worked through in order or used as source material for topic work. Topic areas are defined in the contents list.

The work in Book 2 looks at three aspects of the world:

Part 1   Nature
The first section introduces the nature of the physical world and its plant life. The opening units look at rock and soil – the 'foundations' – and subsequent units explain the variety of plant life which exists on the foundations. They look at plants in detail and show how roots, stems, leaves and flowers have important work to do. The final unit considers the formation of the physical world, the formation of rocks and the development of plant life.

Part 2   Animals
This section starts by getting children to study their immediate locality to discover the many tiny creatures living there. An idea is given of the great variety of soil-dwelling creatures and their importance to the soil. Later units look at larger animals, and a study of bones and joints leads to information on land animals. A look at flight leads to birds, and experiments with water help explain why water creatures are built as they are. The last unit considers the animal kingdom as a whole, and its broad subdivisions.

Part 3   Mysteries
This section looks into some apparent mysteries of the physical world and introduces the idea that they can be explained because they behave according to certain rules. Some of the properties of water, light and electricity are examined. The final unit introduces some of the world's great scientists and the work that they did.

Part 1 Nature

Fossils

Picture 1

basalt, granite, gneiss, limestone, shell fossils

Stones come in many shapes and sizes and colours. Some have a 'picture' of the remains of animals or plants that died millions of years ago. We call it a *fossil*.

From mud into rock

Long ago, the Earth had huge amounts of mud. The mud settled into deep layers which were often covered by water. The top layers of mud pressed down so heavily that they *compressed* the mud at the bottom. This made it hard, and much of it turned into rock.

Making fossils

When animals and plants died, they left bones, shells, stems and leaves in the mud. Their shapes were pressed into the mud. Much later, the mud turned into rock. Soft things rotted away and only their shapes became rock. Hard things like bones often turned into rock themselves.

Sometimes, the hollow shapes left by plants and animals were filled by other materials. These materials became hard and made fossils. You can do an experiment to show how it happened.

> *You will need a margarine tub, plaster of Paris, a tub for mixing the plaster, a mixing spoon or knife, Plasticine, water, a twig*

♥ Experiment: Making a fossil shape

Half-fill the margarine tub with soft Plasticine. Press the twig into it so that it makes a hollow shape. Mix some plaster of Paris into a paste and pour some into the tub. Leave it to harden, then take the tub and Plasticine away (you may have to tear the tub). The plaster will have made a copy of the twig's shape.

Picture 2

Learning from fossils

Fossils are sometimes pushed to the surface by earthquakes. They can be found in some cliffs and quarries where the rock is uncovered. Scientists collect them and study them to discover what plants and animals looked like long ago. This dinosaur skeleton was made by putting fossil bones together.

❦ To write

1 Why did mud turn into rock?
2 How did bones and shells make fossil 'pictures'?
3 How do fossils help scientists?

Picture 3   A dinosaur

❧ More things to do

1 Make a collection of different stones. Find their names.
2 Copy the shapes of other things in plaster. Try leaves and stones. What kind of things make the best copies? Do hard things make better copies than soft ones?

Picture 4   An archaeologist digging up some fossilised bones

Mud and soil

Why do some places get muddier than others? Is mud good or bad for plants? You can find out.

> *You will need four plastic containers with holes in the bottom, a cup, water, gravel, sand, powdered clay and garden soil, three margarine tubs, blotting paper, watercress seeds*

♥ Experiment: How does water soak into the ground?

Take four containers with holes at the bottom. Arrange them so that water can drain through the holes. Picture 1 shows you how. Half-fill one with gravel, one with garden soil, one with sand and one with powdered clay. Pack each one firmly. Then make this chart:

Kind of filling	Number of seconds we counted	
	First try	Second try
gravel		
garden soil		
sand		
clay		

Pour a cupful of water into the container with the gravel. Count the seconds until the water has soaked into the gravel. Write the number of seconds on your chart. Do the same with each container. When you have finished, go round a second time. Use the same amount of water every time.

♥ Experiment: How much water do plants like?

Lay blotting paper inside three containers. Sprinkle watercress seeds into each one. Fill one container with water. Just wet the paper in the second one. Leave the third one dry. Seal their tops on and put them in a warm dark place. Check them each day to see which seeds grow best. Make chart 2 to show your results.

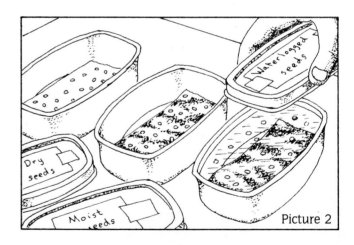

Picture 2

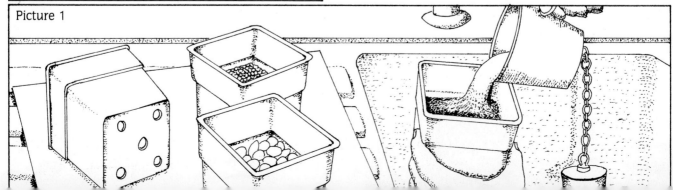

Picture 1

Chart 2	Day 1	Day 2	Day 3	Day 4	Day 5
dry seeds					
wet seeds					
flooded seeds					

Picture 3   What happens when different soils get wet

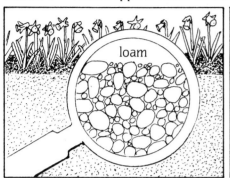 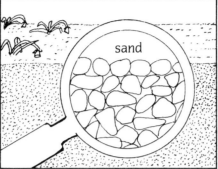 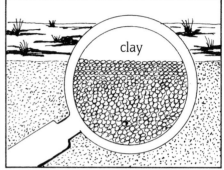

♥ Record

Describe both experiments and draw them. Say if some things let water drain more quickly than others. Does this explain why some places get muddier than others? Say which group of seeds you think will grow best. Say why.

How soil is made

Soil is made from rocks. Rocks are lashed by rain, frozen by ice and scorched by the

Picture 4   Rock splintered by the weather

sun. Tiny grains of rock get blown and washed away. This is called *erosion*. Millions of these grains make up our soil.

Types of soil

Good soil does not drain too quickly or too slowly. Good soil is called *loam*. It stays moist rather than dry or soaking wet. Plants grow well in loam.

Some soils have hard, separate grains. Sand and gravel are like this. The grains leave gaps, and rainwater drains through the gaps before plants can use it. These soils grow poor crops.

Some soils are made of very fine grains which cling together. Clay soils are like this. No gaps are left for water to drain through. They get waterlogged and grow poor crops. The water rots seeds and drowns young plants.

♥ To write

1 How is soil made?
2 Why do some soils get muddier than others?
3 How does too much water hurt plants?

Plants

Picture 1

Rocks and soil are *mineral* things. Plants are vegetable things. What makes plants so different from rocks and soil?

Dig up four small clumps of weeds. Dig deep under their roots and do not damage them. Lay them on a newspaper. Gently pull one clump apart. Divide it into its separate plants. Take one plant and clean the soil from it. Put it on a fresh sheet of paper to study it later.

You will need a trowel, three plastic ice-cream containers, newspaper

♥ Experiment: What do plants need?

Put each of your other three clumps into a container. Pack extra soil around them. You are going to see what happens to these plants when they cannot have certain things. Leave one container without water. Leave one without light, by putting it in a dark place (but do water it each day). Give the third one water and light. Watch them all for a week or two.

♥ Record

Make a careful, full-sized drawing of the plant you cleaned. Describe the plant as if somebody had never seen one before. Say how you started the experiment with clumps of weeds. Say what you expect to happen to them.

Picture 2

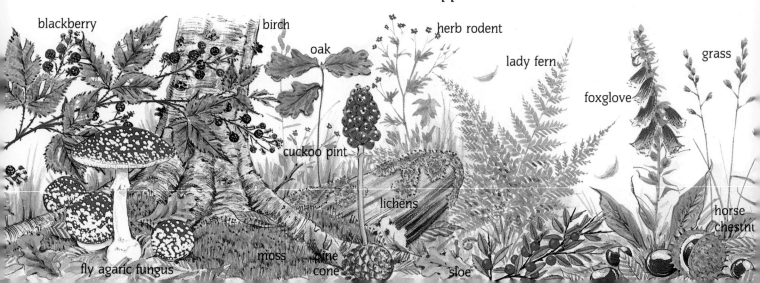

Picture 3 Barley growing on good soil

Picture 4 The soil is poor, plants do not grow well

How plants are like each other

There are many different kinds of plants, from huge trees to things too small to see. But no matter how different they look, all plants are alike in some important ways. All plants grow. They need food to make them grow. This is not the kind of food that we have. The main foods for most plants are sunlight and water. Mineral things do not need food.

Collecting sunlight and water

The earliest plants were tiny things floating on the sea, millions of years ago. They had plenty of light and water all around.

Later, plants began to live on land. They grew roots to hold them in soil and help them get water. They grew leaves to collect sunlight. Plants cannot live without water and sunlight.

Plants need other things. They need air, for example. Vegetable things use light, water and air to make them grow. Plants can make copies of themselves. Minerals do not grow or make copies of themselves.

♥ To write

1 How are all plants alike?
2 What things do plants need to live?
3 How are plants different from minerals?

Picture 5 A simple plant, showing the 'food' it needs

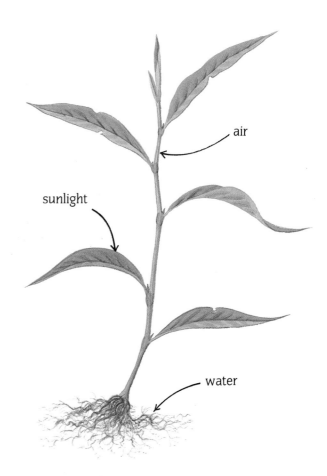

Roots and stems

Picture 1

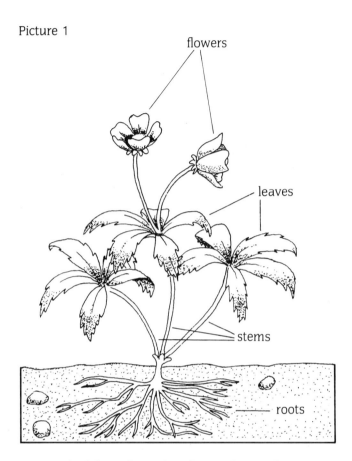

Roots hold a plant firmly in the soil. They support the stem. In turn, the stem supports the leaves. Do the roots and stem have any other work to do?

You will need a carrot, a stick of celery, a knife, two jars, ink or watery paint, salt, water

♥ *Experiment: What do roots do?*
Cut the carrot so that it can stand upright. Cut a deep hollow in the top and put some salt in it. Stand the carrot in water. Look at it during the day. Water will travel through the carrot to the hollow.

♥ *Experiment: What do stems do?*
A stick of celery is a stem. Put a stick of celery upright in a jar of ink or paint. Look at it during the day. The ink will travel up the celery. Cut the stalk in half. See how the ink has travelled up certain parts inside the stem.

Picture 2

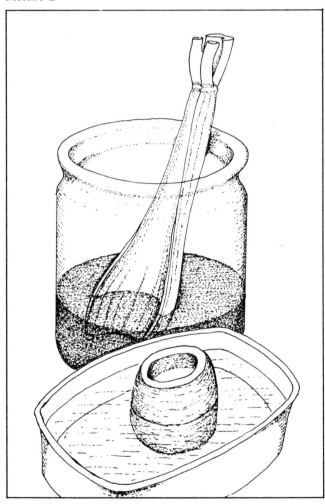

♥ *Record*
Describe and draw your experiments. Say what they have shown you about plants and water. What have you learned about the jobs of roots and stems?

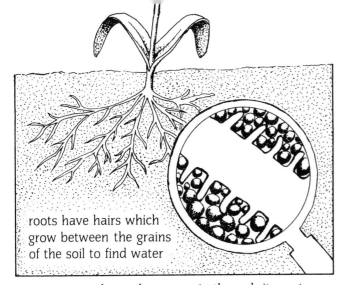

Picture 3   A plant takes water in through its roots

The work of roots

Plants need water to live. They cannot drink like we do. They collect water from the soil. Roots spread into the soil and soak up water through their walls, like the carrot soaks it up. This way of bringing in water is called *osmosis*. The water goes from the roots into the rest of the plant.

The work of the stem

The stem joins the roots of a plant to the leaves. Water travels up the stem through little tubes inside it. These tubes are called *xylem* (say 'zylem'). The celery has xylem tubes inside it. You can see the ink in them when you cut the stalk. The ink travels through them, up the stalk to the leaf. Water travels up tall trees through capillary tubes.

Picture 4   Capillary tubes in a stem

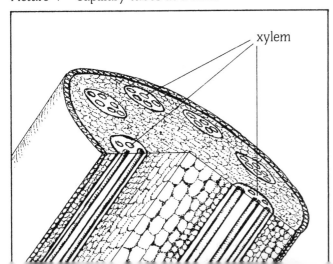

Picture 5   The stem of a tree is called the *trunk*

To write

1   How do roots help plants?
2   How does water travel up stems?
3   What is osmosis?

More things to do

In spring or summer, cut a grass stem in half. Squeeze it and see the *sap* come out. Sap is wetness (*moisture*) on its way up the capillary tubes in the stem.

13

Leaves

Picture 1

Why do leaves turn yellow and fall in autumn? Do trees really need their leaves? Study some leaves for clues.

You will need leaves, paper, paint and brushes, soft pencils

Experiment: Looking for patterns in leaves

Picture 2

Collect some leaves from different trees.
1   Choose a leaf, lay paper on it, and make a rubbing with a pencil. Make a rubbing of both sides. See if this shows a pattern. Do this with each leaf.
2   Choose another leaf. Place it on paper and paint over the leaf and the paper. Take away the leaf and see how its outline is left. Do this with each leaf.
3   Paint each leaf in turn. Lay them on paper and press them gently. See if this leaves a pattern on the paper.

❤ Record

1 Describe your leaves. Mention their colour, size, shape and so on. In which ways are they alike? In which ways are they different?

2 Say if the leaves from the same tree all have the same shape. Use your leaves to find the names of the trees they came from.

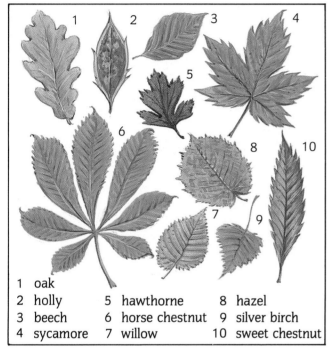

1 oak
2 holly
3 beech
4 sycamore
5 hawthorne
6 horse chestnut
7 willow
8 hazel
9 silver birch
10 sweet chestnut

Picture 3   Different kinds of leaves

How leaves use sunlight

Leaves contain *chlorophyll* (say 'clorofil'), which is what makes plants green. Animals do not have it. Chlorophyll lets leaves use sunlight to help them grow.

The leaves take in air (the part of air called *carbon dioxide*). They also get water from the roots. Chlorophyll lets the leaves make a kind of sugar from sunlight, air and water. This sugar is called *starch*, which is the food that plants need.

Leaves give out *oxygen* and spare water. Oxygen is an important part of air, because animals cannot live without it.

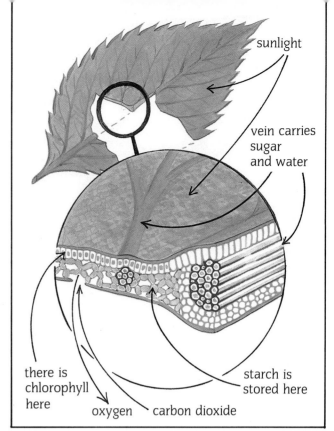

Picture 4   How a leaf works

A pattern made of veins

Inside large leaves are thin tubes called veins. These veins are passageways for starch to move through. They also support the leaves. The veins are arranged in a pattern. Leaf prints show up these patterns of veins.

Autumn

Leaves change colour in autumn because they lose chlorophyll. In autumn, the trees seal off their leaves so that nothing can go in or out of them. This makes the chlorophyll break down. The leaves change colour, die and fall away. Plants do not grow in winter, and do not need leaves.

❤ To write

1 Why is chlorophyll important?
2 Why do leaves have veins inside them?
3 Why do trees not have leaves in winter?

Flowers at work

Picture 1

Flower-heads shake in the wind. Is this to make them look attractive, or is there another reason? Do the experiment to find out.

 You will need thread, masking tape, scissors, Plasticine, a pencil, salt, coarse and fine sand, gravel, a piece of paper

♥ Experiment: Scattering grains by wind

Clear your desk. Tape lengths of thread across it at these distances from one end: 25cm, 50cm, 75cm and 100cm.

Make a model flower from Plasticine and a pencil. Put into the top of it some grains of salt, some small grains of sand, some large grains of sand and some small pieces of gravel. Roll the piece of paper up into a tube and blow through it at the flower. Make a wind that will blow the grains along the desk.

See where the heavy grains land. See where the light grains land. See where the others land. Record your results, and then try the experiment again.

Picture 2

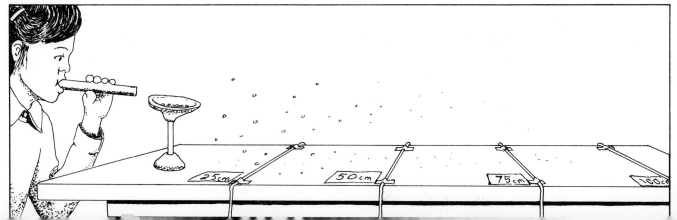

♥ Record

1 Make this chart to show your results:

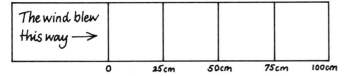

2 Describe what you have done. Say what you found about how a wind spreads light and heavy grains. Make a rule to say how wind treats light grains and heavy grains differently.

What is pollen?

Flowers produce tiny grains called *pollen*. They also grow parts called *ovules*. To start a new plant, pollen from one flower has to enter an ovule in another. So pollen needs to travel through the air.

Picture 3   A simple flower

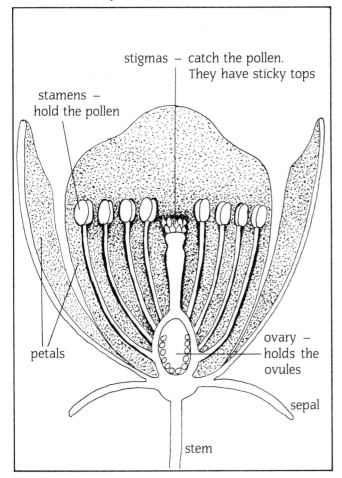

Pollen is spread by the wind

Sometimes, wind blows the pollen. You cannot see the pollen because it is too small, and the wind can blow it a long way because it is so light. Your experiment showed how the lightest things are blown furthest by the wind.

When the pollen falls on other flowers, some of it will enter ovules. This is called *fertilising* the ovules.

Pollen is spread by insects

Insects carry pollen from one flower to another. You may have seen bees feeding on flowers. They pick up pollen on their bodies, and take it with them when they fly to other flowers. Some of the pollen fertilises the ovules. Then seeds begin to grow in the ovules.

Picture 4   A honey bee feeding on a flower

♥ To write

1 What is pollen?
2 How does pollen get from flower to flower?
3 What has to happen to start a new plant?

Summary: The world

How the mineral world was made

Picture 1

Picture 2

The world was formed many millions of years ago. It may have started as a huge cloud of dust and gas in space. This cloud spun round and round (*rotated*) and grew smaller. As the parts came closer together, they got hotter. The cloud became more solid and grew very hot indeed, melting the different *elements* or materials. The elements mixed together, and the heavy ones sank towards the centre of the Earth. Some rose towards the surface, and some became the waters of the sea.

The making of rock

The Earth grew cooler and the elements on the surface became cold. They became solid rock. There were different kinds of elements, which cooled at different speeds. They made different kinds of rock.

Some rocks were made in another way. Chalk began as layers of tiny dead creatures which lay at the bottom of the sea. The top layers pressed down for millions of years, and the crushed lower layers became chalk.

Coal was made from the plants of early forests. After the plants died, they were covered by others which pressed down on them for millions of years and crushed them. They turned into coal.

What the world is like today

The world is almost all rock, thinly covered by soil. The rock near the surface of the Earth is solid. This is the Earth's *crust*.

Deeper down, the rock is still hot. It is liquid, and we say it is *molten*.

Picture 3 A 'slice' through the Earth

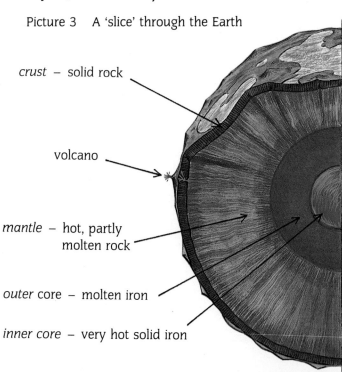

crust – solid rock

volcano

mantle – hot, partly molten rock

outer core – molten iron

inner core – very hot solid iron

The first plants

The first plants that appeared on Earth were very simple ones, called *algae*. The first algae floated in the sea. They had chlorophyll in them, and used the water and sunlight to make oxygen. Over millions of years, they gradually made the air that we breathe. After more millions of years, algae and other simple plants began to grow on land.

The development of plants

As time passed, plants grew larger and more complicated. Mosses and ferns began to grow. Later, large plants such as trees developed. All the plants in the world have developed from the simple algae that lived in the oceans long ago.

Picture 4   How plants developed

Part 2 Animals

An animal hunt

You are surrounded by wild animals, which are easy to find if you search for them.

♥ *Experiment: Searching for soil creatures*

You will need a trowel and a copy of this chart

Type of animal	Where we found it					
	In the soil	On soil, stone or wood	On a plant	In water	Under cover	Anywhere else
snail						
slug						
worm						
beetle						
earwig						
woodlouse						
spider						
centipede						
millipede						
ant						
others						

Search in the grounds of your school. Look in open land and overgrown places, on paths and under stones. Use a trowel to dig in the soil. Each time you spot a creature, tick the chart to show where it was. Be careful not to harm any of the animals, and put back all stones, wood or soil that you move.

♥ *Record*
1. Describe your search for soil creatures.
2. Write a sentence to describe each creature you found.
3. Write some ways in which these creatures are like each other.
4. Write some ways in which they are different from each other.
5. Say how animals are different from plants.

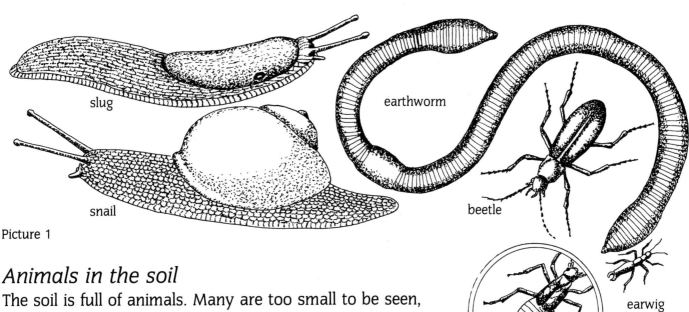

Picture 1

Animals in the soil

The soil is full of animals. Many are too small to be seen, but even the smallest are important.

Slugs and snails eat dead creatures and plants. They are food for birds.

Earthworms burrow in the soil. They improve it by making it looser and letting in air. They are food for birds, beetles and centipedes.

There are many kinds of *beetles*. Some eat wood, some eat plants, some eat other creatures. Some eat dung.

Earwigs eat dead creatures, leaves and roots.

Woodlice eat decaying things and live in damp places. Their favourite foods are dead wood and leaves.

Many spiders weave webs in which to catch flies and other insects for food.

Centipedes live in the soil and under stones. They eat other living creatures.

Millipedes feed mostly on rotting things. They help to turn dead things back into useful things for the soil.

Ants live in large groups and burrow nests in the soil. They will eat almost anything.

These creatures are important for the soil. They break down rotting plants and dead animals, turning them into things which are good for the soil. They turn them into food for fresh plants.

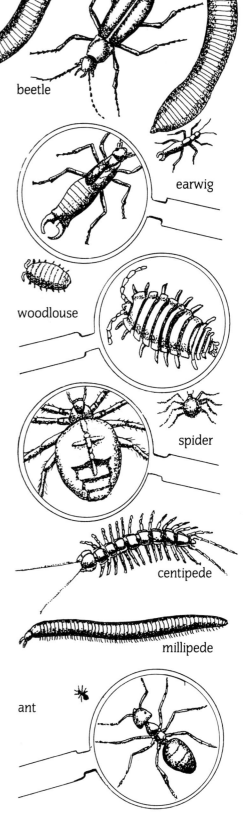

❤ To write

1 How do earthworms improve the soil?
2 What happens to plants and animals when they die?
3 Why are soil creatures important?

Animal traps

Soil creatures like to stay hidden, but you can catch them in a pitfall trap. You can then study them and let them go unharmed.

 You will need two jars, cardboard, scissors, small pieces of wood, small pieces of meat, fruit and leaves

♥ Experiment: Making pitfall traps

Picture 1

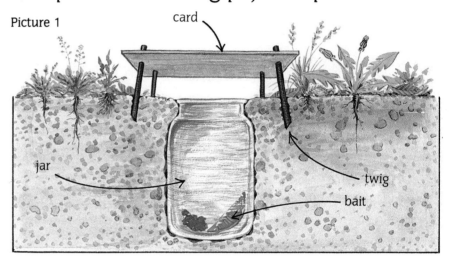

Sink a jar into the soil. Put a square of card over the top to keep out rain. Use bits of wood to hold the card above the ground. Put two pitfall traps in the school grounds. Think of good places to put them and tell your teacher where they are. Bait one of the traps with food. What things would be best to put in? Try leaves, meat and fruit. Look at your traps each morning and evening to see what you have caught. Let the creatures free each time you check the traps. Never leave creatures in a trap, and be careful not to hurt them.

At the end of a week, put the traps away.

♥ Record
Copy this chart and fill it in, then answer the questions.

	\multicolumn{6}{c}{What we found in the traps}					
	first morning	first evening	second morning	second evening	third morning	third evening
trap 1 (no bait)						
trap 2 (baited)						

Picture 2 Foxes are carnivores

Picture 3 Sheep are herbivores

1 Do you catch more creatures at night or in daytime? Why?

2 Does a trap with food catch more creatures than an empty trap? If it does, why?

Different ways of feeding

You will probably catch more creatures at night. The baited traps may catch more creatures. This is because most soil creatures look for food at night. Soil creatures eat living and dead plants and animals. Some eat only plants, and are called *herbivores*. Some eat only other creatures, and are called *carnivores*. Some creatures eat plants and animals, and are called *omnivores*.

Life starts with the sun

All animals depend on plants for food. Even carnivores depend on plants, because carnivores eat animals which live on plants. Plants can grow only where there is sunshine to give light and warmth. Because plants depend on the sun, animals depend on it too. The picture shows how the sun gives life to everything on Earth:

The sun gives heat and light. Plants turn sunlight into food. Herbivores eat the plants.

Omnivores eat the herbivores.

Carnivores eat the omnivores.

Picture 4

The sun is the *source* of all life.

♥To write

1 What is a herbivore?
2 Why could there be no life without the sun?
3 What do most soil creatures do at night?

Bones

Picture 1

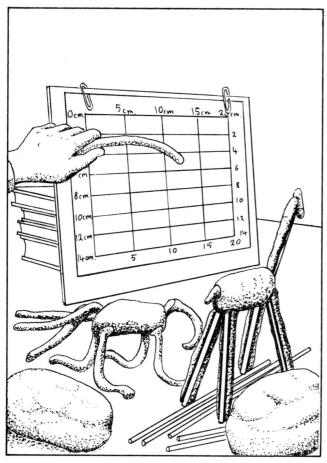

Soil creatures have no bones, yet large animals do. Why do large animals need bones?

You will need Plasticine, drinking straws, scissors, cardboard

♥ Experiment: Inventing an animal

Invent an animal. Join thin lengths of Plasticine to make legs and a body. Make the animal as tall as possible.

You will soon find a problem. The taller and thinner the animal gets, the more it will droop. It needs support.

Cut lengths of plastic drinking straw to support the Plasticine. See how tall you can make the animal now.

♥ Testing stiffeners

Draw a chart like the one in the picture, and prop it up. Hold lengths of Plasticine beside it, and mark on the chart how they droop. Then stiffen the Plasticine with straws. See how much less they droop then. Mark the differences on the chart.

♥ Record

1 Describe how you made a model animal. Draw it, and say how stiffeners helped it.

2 Describe how you held lengths of Plasticine next to the chart to see how they bent. Say what difference stiffeners made.

Picture 2   Animals and their skeletons. Can you name them?

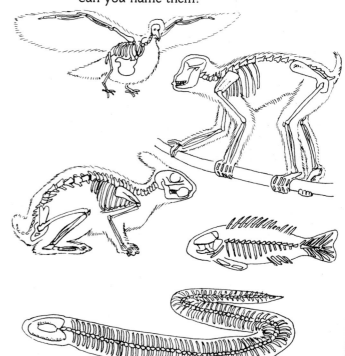

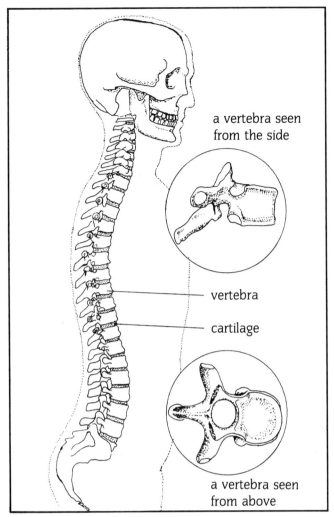

Picture 3 How the backbone is made up

Picture 4 Why giraffes need their skeletons!

The skeleton

Bones make a framework to support the body. This framework is called a skeleton. The main part of the skeleton is the backbone, which is made up of lots of bones. Each bone is called a *vertebra*. Animals with backbones are called *vertebrates*. You are a vertebrate.

The bumps down the middle of your back are the bones of your backbone.

Cartilages are spongy parts between the bones. They let the backbone bend easily.

Arms and legs are made with long bones with joints between them. The joints allow the arms and legs to bend.

Bones are strong. To stop them from being too heavy, they are hollow.

When an animal dies, the soft parts of its body rot away. Its bones do not rot.

To write

1 Why do animals need bones?
2 Why do backbones have many short bones instead of one long one?
3 What kinds of animals do not have bones?

Something to do

Next time you have roast chicken at home, look at its bones.

How bones move

Our bodies are held up by a strong framework of bones which is able to move. Movement is made possible by joints.

> You will need card, scissors, string, a paper rivet, half of a rubber ball, Plasticine, washing-up liquid, a pencil

♥ Experiment: Finding out how a hinge joint works

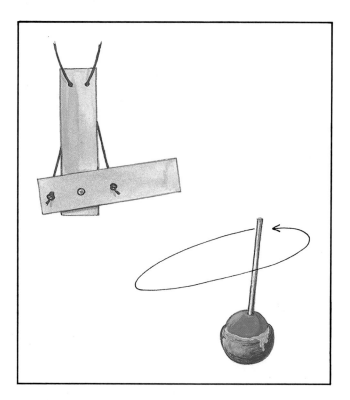

Cut two strips of thick cardboard 5 cm by 20 cm. These are bones. Make holes in them and join them as the picture shows. Use pieces of string to act as muscles. Hold the upright strip and pull on the strings to make the other strip move.

You have made a model of how your elbow works.

♥ Experiment: Finding out how a ball and socket joint works

Take one half of a rubber ball and make a ball of Plasticine to fit into it. Push a pencil into the Plasticine. Smear washing-up liquid between the Plasticine and the inside of the ball.

The pencil swings around easily in all directions. You have made a model of how your shoulder joint works.

♥ Record

1 Describe two joints. Say how they are alike and how they are different.

2 How many joints can you count in your body? Which are hinges and which are ball and sockets? Make a list of what you find.

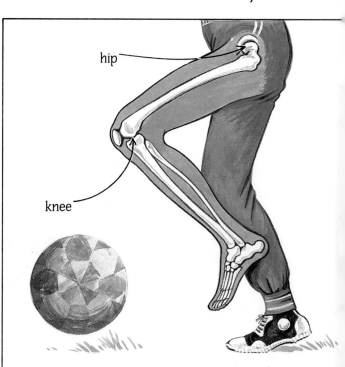

Picture 2 The shoulder and hip are *ball and socket* joints. The elbow and knee are *hinge* joints.

Joints

Our bodies are able to move because our bones have joints between them. Most are simple hinge joints, but they allow only up and down movement. Shoulders and hips have ball and socket joints, which allow much more movement. You can test this by swinging your arm or leg around.

Muscles

Joints are worked by muscles, which are fixed to bones like the strings on your first model. When muscles shorten themselves, they pull on the bones and make them move.

Animals' bodies can make many movements because they have many muscles. A human being has over six hundred, most of which are small and work things like eyes and mouth. The big muscles which work arms and legs can be seen and felt.

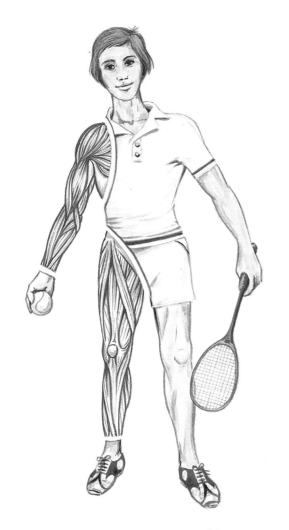

Picture 3 The muscles of the arm and leg

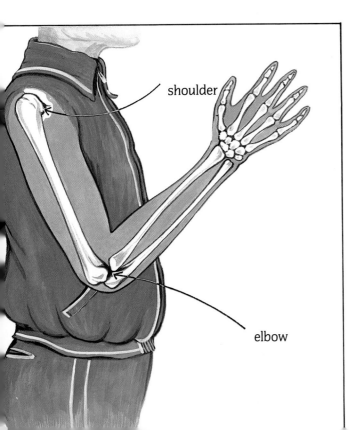

Picture 4 A muscle man!

♥ To write

1 Why do skeletons have joints?
2 How is a shoulder joint different from an elbow joint?
3 How do we make our bodies move?

Flying

Picture 1 Unstable flight

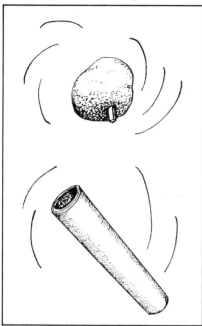

Picture 2 Stable flight

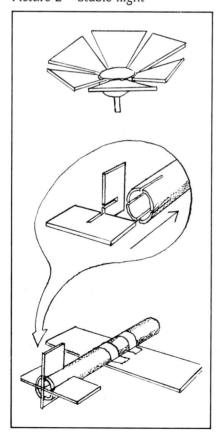

Picture 3 A herring gull

Birds are flying animals. The air gives them support, and allows them to fly in a way which only the insects can copy.

 You will need Plasticine, a matchstick, cardboard, scissors, paper, sellotape

♥*Experiment: What is unstable flight?*

Push a matchstick part-way into a small lump of Plasticine. Toss the lump into the air and let it fall to the ground. Do this several times. How many times does it fall on the matchstick? It will not always land on the matchstick, because the lump tumbles over and over in the air. It is *unstable* when it flies.

Make a tube from a piece of paper. Mark it on one side and throw it into the air. Try to make it fly. Is it stable or unstable? How often does it land with the mark on top?

♥*Experiment: What is stable flight?*

Cut six vanes from stiff cardboard. Press them into the Plasticine as the picture shows. Now toss the Plasticine again and see if it lands on the matchstick.

Fix more vanes to the tube. Fly it again, and see if it is stable or unstable. How often does it land with the mark on top?

♥ Record

1 Copy this chart and fill it in:

	First throw	Second throw	Third throw	Fourth throw
Plasticine on its own				
Plasticine with vanes				
Tube on its own				
Tube with vanes				

2 Describe your experiments. Draw the things you used.

3 What is unstable flight? How can you make it stable?

Feathers and wings

Things fly better when they are shaped in a certain way. The pictures show how air streams past a moving object. If the air

Picture 4   Why some things fly better than others

streams past smoothly, the object flies steadily. *Stabilisers* make the air stream past more smoothly. They keep objects steady and let them fly better.

Birds' wings are stabilisers made almost completely of feathers. A feather (shown in close-up) is light, strong and flexible. This means that it bends without snapping. The wing of a bird is almost all feathers. A bird flaps and twists its wings in a certain way as it moves through the air, which gives it *lift* to fly. Feathers help a bird to have a streamlined shape, so that it slips through air easily.

Feathers trap warm air, which helps to keep the bird warm.

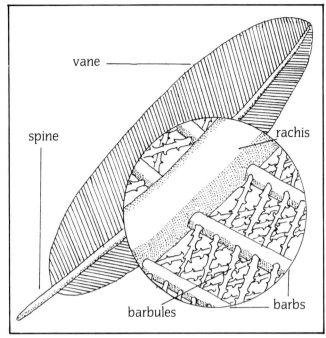

Picture 5   How a feather is made up

♥ To write

1 How do vanes and feathers help a thing to fly?

2 What does a bird do with its wings to make itself fly?

3 Do feathers help a bird in any other way?

Life in water

Simple experiments can show why water animals are shaped differently from land animals.

> You will need a bowl, water, a plastic straw, Plasticine, a 1kg weight, a spring balance

♥ Experiment: Looking at underwater drag

Put a bowl of water on your desk. Press a ball of soft Plasticine around one end of a plastic drinking straw. Press the Plasticine into a long, thin shape. Put the Plasticine into the water and use the straw to move it around. See how easily it slips through the water.

Then press the Plasticine into a wide, flat shape. See if this moves as easily. Try it with several different shapes. Which moves most smoothly?

Picture 1

♥ Experiment: Measuring underwater lift

Tie a 1kg weight to a spring balance. Hang the weight from the balance and read the scale. Now lower the weight into water. Read the scale again, and see if it has changed.

Picture 2

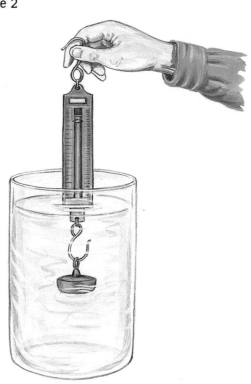

♥ Record

1 Describe and draw your first experiment. Say what kind of shape moves best through water.

2 Describe and draw your second experiment. Say what happens to the weight of things in water.

3 Can you explain why sea creatures have the shapes they do?

Streamlining

Your first experiment showed that long, thin shapes slip easily through water. Wide, flat ones do not. Water resists a wide shape because it cannot stream past it. Thin, smooth shapes let water slip past easily. Such shapes are *streamlined*. Fish are streamlined.

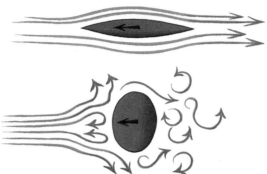

Picture 3
Why fish need to be streamlined

Upthrust

Fish float because the water helps to hold them up. The water supports them just like it supported the weight and made it pull less heavily on the spring balance. This support is called *upthrust*. It is the main reason fish do not have legs to help them move. Instead they float in the water and they move by flexing their bodies and fins to drive themselves through the water.

Fish do have a backbone such as land animals have, and a framework of bones.

Sea creatures grow larger in water than they could on land, because the water supports their bodies. The jellyfish is an example. It floats in water, but it would collapse on land.

Picture 4   A stickleback. Look at its shape

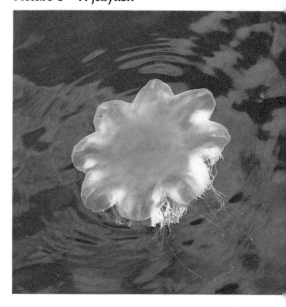

Picture 5   A jellyfish

❤ To write

1 How do the shapes of sea creatures help them?
2 Why do sea creatures not need a strong framework of bones?
3 What does 'streamlined' mean?

➤ Something to do

Next time you go swimming, float flat and feel how the water supports you. Notice how heavy you feel when you get out.

Summary: The animal world

Picture 1
The first animals were simple creatures in the early seas.

Some of them developed into fish.

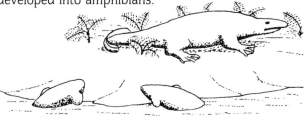

Some fish learned to spend time on land. They developed into amphibians.

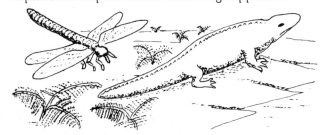

Reptiles developed. Insects with wings appeared.

Some reptiles developed wings. The first small mammals appeared.

The large reptiles died out. Insects, birds and mammals spread.

Evolution

There have been animals on Earth for millions of years, but when living things first began they were different from those we see today. Living things began in the seas, which once covered most of the world. The first animals were simple, tiny creatures that floated in the water. They changed slowly, or *evolved*. Some evolved to live in warm water, some to live in cold water. Others evolved to live in deep or shallow water.

Life on the land

Some changed so that they could live out of water, and learned to move by wriggling on the ground. Arms and legs developed. Others grew fur or scales. Some learned to fly, and grew feathers. Slowly, animals evolved into the thousands of different types which make up the *kingdom* of animals.

The animal kingdom

The animal kingdom contains two main types of animal: those which have backbones, and those which do not. Those which have backbones are called vertebrates. Those which do not are called *invertebrates*.

Invertebrates Invertebrates include the simplest animals such as worms, slugs and snails. Squids, octupuses, jellyfish and crabs are invertebrates.

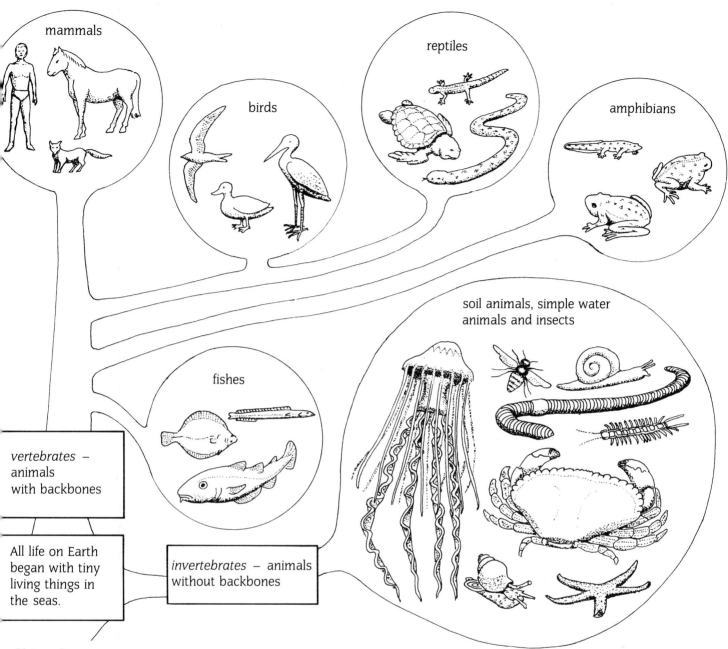

Picture 2

One of the largest groups of invertebrates are the insects. They are animals which have jointed legs and bodies, with a tough outer covering. Examples of insects are beetles, ants and flies.

Vertebrates Vertebrates include *mammals*. Mammals are animals which have warm blood, hair or fur, and which make milk for their babies. Human beings are mammals.

Men and women appeared on the Earth a long time after life first started. The first men and women were smaller than we are, and were probably covered with hair. Scientists are still trying to discover more about the earliest people.

Reptiles such as crocodiles and lizards are vertebrates, as are the *amphibians*, such as frogs and newts, which live in and out of water. Birds are vertebrates, and so are most kinds of fish.

Part 3 Mysteries

Walking on water

Picture 1 A raft spider

When things float, part of them usually lies under the water. Some creatures, like the one in the picture, are so light that they are able to walk on the water, without going through the surface. Is there something about the surface of water that helps them to do this?

> You will need a bowl, a paper clip, blotting paper, scissors, a jar, water

♥ Experiment: To show the surface of water

Fill a jar completely with water. See if the surface of the water is exactly level with the rim, or if it is higher than the rim.

♥ Experiment: To show how strong this surface is

Drop a paper clip into the water. It sinks. Can you make it float?

Cut out a small piece of blotting paper and lay the clip on it. Lay the blotting paper on the water. When it is full of water, it will sink to the bottom. If it does not, tap it gently. (Do not disturb the paper clip.) See if you can get the paper clip to float. Try the experiment a few times.

Picture 2
The water curves above the top of the jar

♥ Record

Describe what you did, and say what happened each time. Use pictures to make it clear.

Surface tension

Imagine that you built a model castle out of thousands of toy bricks. If people saw it from far away, they would not notice the bricks. The bricks would be too small for them to see.

Everything in the world is made up of tiny 'building bricks' called *molecules*. They are far too small to be seen. Water is made up of millions of molecules.

Water molecules normally cling together. At the surface, they cling with extra strength. Nobody knows why. This extra strength is called *surface tension*. It means that water at the surface can hold things up which would normally sink. It is surface tension that keeps water higher than the rim of a jar, and lets pond skaters walk on the surface.

Picture 3

Picture 4 How molecules stick together. At the surface they hold on very tight

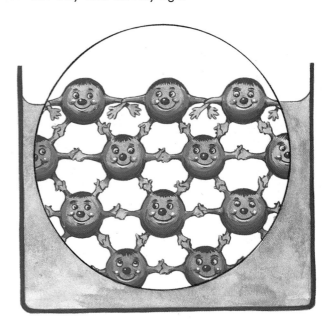

♥ To write

1. What is water made of?
2. Why can you not see molecules?
3. How can a paper clip float on water?

On the level

Picture 1
An aquarium

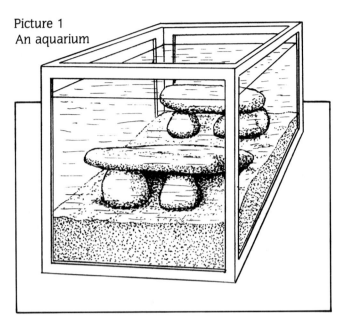

Is it possible to empty this aquarium without disturbing anything?

Picture 2

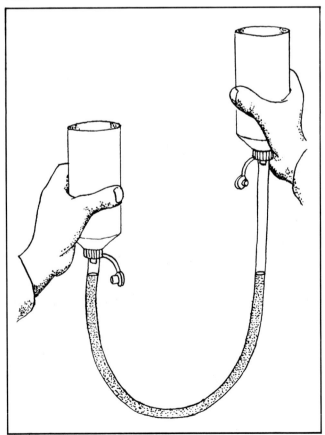

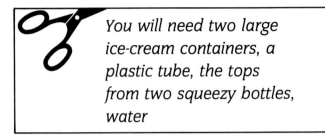

You will need two large ice-cream containers, a plastic tube, the tops from two squeezy bottles, water

♥ Experiment: To show how water keeps one level

Fit a plastic tube to two tops cut from squeezy bottles. Put some water into the tube. See how the water levels stay in line with each other. Lift one side higher, and then put it lower. See how the water levels stay in line.

Picture 3

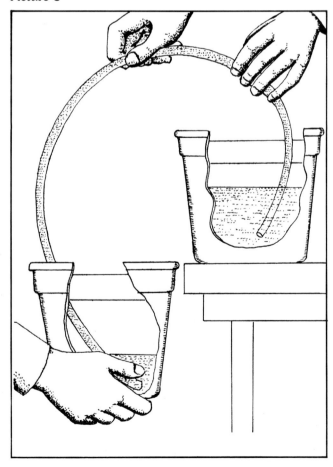

36

♥ Experiment: Moving water from one level to another

Fill a container with water. Hold an empty container lower down. Fill a plastic tube with water so that there is *no* air left in it. Hold a finger over each end of the tube. Put one end into the full container. Put the other end into the empty one. Let go of the ends. Water will flow from one container to the other.

See if you can make the water flow back again. Can you stop it from flowing altogether?

♥ Record

Write about your experiments, and draw them. Say what you learned about water levels. Say how you can make water move.

Water wants to be level

Water never stays high when it can move downwards. Imagine two lakes with different water levels. If they are joined, the water will mix. It will make a new level. This is what your second experiment showed. The tube joined the containers so that water could flow between them, and it tried to make one level. It flowed from the high container into the low one. When you lifted the low container high, the water flowed back. It was still trying to get level. When it was level, it stopped moving.

A tube used to make water flow like this is called a *siphon*. A siphon can empty an aquarium without disturbing anything.

Sea level

The water in all the seas of the world tries to make one level. This is called sea level. Most land, but not all, is above sea level.

♥ To write

1 What does water always try to do?
2 What happens when two different levels of water join?
3 What is a siphon, and how does it work?

Picture 4 The water moves to make one level

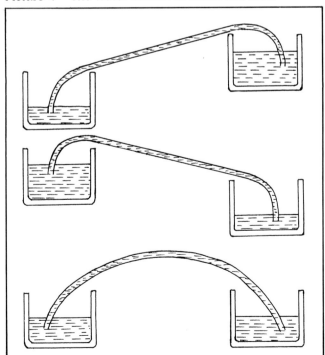

Picture 5
A waterfall – the water is trying to make one level

Can you believe your eyes?

If you put your arm in water, it will seem to bend. When you take it out, it looks straight again. What makes this happen?

You will need a bowl, a stone, a ruler, water, card, a toy, Plasticine, scissors, chalk

♥Experiment: To show that light travels in a straight line

Take three sheets of card and cut a slit in each one. Stand a toy on your desk. Stand the cards between your eyes and the toy. Move them around until you can see the toy. When you can see it, make a chalk mark where each slit is on the desk. Take away the cards. See how the markers always make a straight line. (Clean the chalk marks off the desk.)

Picture 1 Does light travel in straight lines?

♥Experiment: To show how water affects light

Put a stone on the bottom of a bowl. Lean a ruler against the side of the bowl. Fill the bowl with water. Does the stone seem closer? Does the ruler seem to change?

38

Picture 2 The mystery of light and water

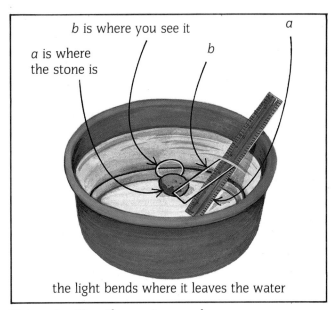

Picture 3 How the mystery works

♥Record
Describe your experiments, and draw pictures of them. Say how water changes the way things look. Water does not change the things themselves, so what does it change?

Light lets us see things
We see things because of light, which travels from them to our eyes. Our eyes tell our brains that the things are there. The first experiment showed that light travels in a straight line. Because it travels straight, light shows exactly where things are.

How water affects light
Light travels well through air, but not so well through water. When light goes in or out of water, it changes direction, like a stone thrown into a pond. It goes into the water at an angle, then sinks at a different angle.

Water *refracts* or bends light, which is why things look different under water.

How things look different under water
When the bowl was empty, you saw where the stone was. When it was full of water, the water bent the light going from the stone to your eyes. Your eyes were fooled into thinking that the stone was closer than it was. When you put your arm in water, your eyes are fooled in the same way.

♥To write
1 In what direction does light travel?
2 How does water affect light?
3 What happens to light when it is refracted?

Picture 4 Light cannot travel far underwater

39

Moonlight

The moon shines brightly at night but not in the daytime. Why?

The moon sometimes looks round, but sometimes has other shapes. Why?

You will need a large ball, a small ball, string, tape for joining string to the balls, scissors, a torch

❤ Experiment: A model Earth and moon

Hang a large ball from a string. Imagine that this is the Earth. Hang a small ball from a string. Imagine that this is the moon. Take the moon in a circle round the Earth. This is how the moon goes round the Earth.

Picture 1

Take your models to a dark place. Shine a strong light at them. The light stands for the sun, which makes light and shadow on the Earth and moon. Circle the moon round the Earth again and watch the light and dark sides of the moon. Think how the moon will look from the Earth. Sometimes only the light part will be seen, and only the dark part at other times. Sometimes, a mixture of light and dark is seen.

❤ Record

Describe and draw the experiment. Draw circles and shade them to show how the moon appears from the Earth. Does the moon really change its shape?

Picture 2 The moon can light up the night

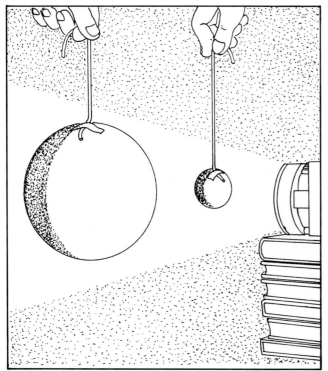

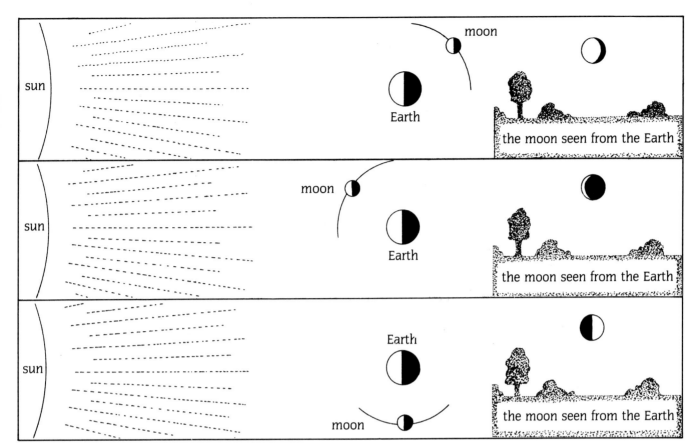

Picture 3 How light travels from the sun and moon to the Earth (Not drawn to scale.)

Day and night

The moon does not make its own light. When it shines, it is only reflecting light from the sun. The sun gives day and night to the Earth and the moon, making one half light while the other is in shadow.

The moving moon

When we see the daytime side of the moon, we call it a full moon. Sometimes we see part of the daytime side and part of the night-time side, or only the night-time side. The moon is moving round the Earth all the time, and it seems to change shape. It shows different amounts of its bright side and its dark side. It does not really change its shape.

The moon seems dull in daytime because the sun makes the whole sky bright.

❤ To write

1 How does the moon make moonlight?
2 Why does moonlight seem brighter at night?
3 Why does the moon seem to change its shape?

Rainbows

You will need a glass prism, a small mirror, a small bowl, water, a torch

♥ Experiment: Using glass to make colours

In a dark place, shine a torch through a glass *prism*. You will see that the prism splits the beam of light into different directions. It refracts the light. If you hold the prism to your eye, you will see that it surrounds things with colours. These are the colours of the rainbow.

♥ Experiment: Making rainbow colours

Lower a small mirror into a bowl full of water. Tilt the mirror backwards, and look at your reflection in it. See if your reflection has extra colours round the edges. Tilt the mirror backwards and forwards, and from side to side. Can you see extra colour?

Picture 1

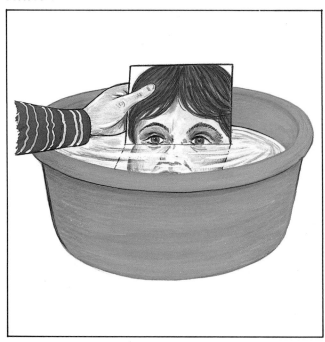

Picture 2   A double rainbow. How many colours can you see? (There should be seven different colours)

Picture 3   Isaac Newton experimenting with light.
A prism can split light into colours

♥Record
Describe and draw the two experiments. Say if you managed to see extra colours. What colours were they? What do you think caused them to appear?

Picture 4   A spectrum in a hosepipe spray

Colours from sunlight
Light is refracted when it goes through glass or water. If it is refracted enough, it splits into different colours. The man who discovered this was Isaac Newton, in 1666, when he shone light through a prism on to a screen. The light split up into the colours of the rainbow. Scientists call them the *spectrum*.

You saw the colours of the spectrum when you looked through the prism. The mirror in water made the same thing happen. The two experiments refracted light enough to make some of it split into colours.

How rainbows are made
After rain, tiny drops of water stay in the air. They are not heavy enough to fall. Sunlight shines through them and they refract it. The light is split into the colours, which makes a rainbow appear in the sky.

♥To write
1 How can coloured light be made from ordinary light?
2 What is the spectrum?
3 How are rainbows made?

Lightning and other sparks

Picture 1

Picture 2

You will need a balloon, a plastic comb, wool or nylon cloth, paper, scissors

♥ Experiment: Making an electric charge

1 Blow up a balloon and tie it. Rub it against something woollen, such as a pullover. Hold the balloon against something and see if it clings to it. Rub the balloon again and hold it close to your ear. Can you hear crackling noises?

2 Take a plastic comb and some wool or nylon. Cut a sheet of paper into small pieces. Rub the comb strongly and quickly with the cloth. Hold it near the paper, and see if anything happens. Try both experiments several times.

♥ Record

Describe and draw pictures to show what you did and what happened. Say how rubbing changed the balloon and comb. What could they do? Did you hear anything?

Static electricity

When you rubbed the balloon and comb, you made a charge of electricity. Made like this, it is called *static electricity*.

When something is charged with static electricity, it can pull at things. This is because the electricity tries to go from one thing to another. When it has gone from one thing to another, the pull stops.

Paper falls off a charged comb after a while. When the electrical charge has gone, we say that it has discharged.

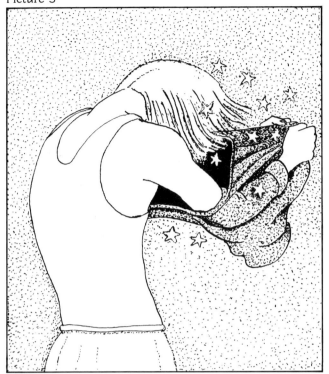
Picture 3

Thunder and lightning

Lightning is caused by static electricity. Some kinds of weather give the air an electric charge, which discharges when it is strong enough. A great spark flashes between the air and the ground. The noise it makes is what we call thunder.

Lightning is a huge discharge of static electricity.

Picture 4 Thunder and lightning often happen together

You can feel and hear static electricity. Sometimes, when you undress, you hear crackling, and your skin feels strange. This happens especially when you take off nylon clothes. These clothes build up an electrical charge during the day, and when you take them off, it is discharged. This is what you can hear and feel. If it is dark, you may see sparks.

♥ To write

1 What can rubbing certain materials build up?
2 Why do our clothes crackle sometimes when we undress?
3 What is lightning?

Picture 5
Lightning – a dramatic discharge of static electricity

Summary: Explorers of mysteries

People have always explored mysteries in science, and have tried to work out how and why things happen. Here are some of them:

Archimedes

Archimedes lived nearly three hundred years before Jesus. He lived in Greece. One of his discoveries was that when something sinks in water, it pushes the water level up. Legend says that he noticed this when he got into a bath (check this when you have a bath). He invented a machine to lift water from a low level to a high one. You can see one in Picture 1. As the screw is turned, it carries water up the tube.

Picture 1   Archimedes' screw

Picture 2

New words

algae tiny plants that have no roots, stem or leaves
amphibians animals that can live on land and in water

carnivore an animal that eats other animals
cartilage tough, bendy stuff that lies between bones; it is like the gristle in meat
chlorophyll the green stuff in plants that lets them make sugar from sunlight, air and water
compress to press something until it is hard and solid

elements the things from which all the rocks, water and air of the world have been made
erosion the wearing away of rocks by the weather
evolve to change gradually

fertilise to start the growth of a new plant or animal
fossil a copy in stone of a plant or animal that lived long ago

herbivore an animal that eats plants

invertebrate an animal that does not have a backbone

lift a force which pushes upward
loam a good soil, not too sandy or clayey

mammal an animal that has warm blood, makes milk for its babies and has hair on its body
minerals things which are not alive, especially solid things such as rock
moisture wetness
molecules the tiny pieces of which everything is built
molten a melted solid

omnivore an animal that eats plants and animals
osmosis the way that water enters plants through their walls
ovule the part of a plant that can be fertilised by pollen to start a new plant
oxygen part of the air that we breathe

pollen grains made in flowers to fertilise ovules in other flowers
prism a specially shaped piece of glass or plastic

refract to bend light away from its usual straight line
reptile a cold-blooded animal that lays eggs, has a backbone and has scales or plates on its outside

sap the wetness inside a plant made of water and the foods that the plant needs
siphon a tube used to move water out of a container to a lower level
spectrum the colours that can be made by refracting light
stabiliser a shape fixed to an object to make it fly smoothly
static electricity electricity made by rubbing things together
streamlined shaped so that water or air slips past easily
sugar a food plants make from water, air and light
surface tension the strength that water has at its surface

unstable not well balanced
upthrust the support that water gives to make things less heavy

vertebra one of the bones in a backbone
vertebrate an animal that has a backbone
xylem thin tubes that carry water up a stem

Copernicus

Copernicus was born in Poland in 1473. He was interested in the sky. People used to believe that the sun and stars travelled round the Earth. Copernicus watched carefully to see how the stars and planets moved in the sky, and worked out that the Earth travels round the sun. He realised that the stars do not travel round the Earth.

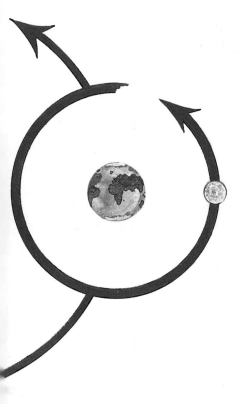

Galileo Galilei

Galileo Galilei carried on the work of Copernicus. He was born in Italy in 1564 and was the first man to study the sky through a telescope. He agreed that the Earth and the other planets go round the sun, and discovered that the moon has a rough surface and not a smooth one. He showed that the moon gives light by reflecting sunlight.

Isaac Newton

Isaac Newton, a British scientist who lived in the seventeenth century, experimented with light. He found that light can be split into different colours, and did this by shining light through a glass prism. By using two prisms, he was able to make light look normal again.

William Gilbert

William Gilbert lived in England four hundred years ago. He saw that there was a pulling force between certain things, and called it an electric force.

Marie Curie

Marie Curie, a Polish scientist, studied radioactive materials. In 1902 she and her husband discovered radium. She won the Nobel Prize twice.

Picture 3
Marie Curie in her laboratory